The
CAT LOVER'S
Compendium

First Skyhorse Publishing edition published 2014 by arrangement with Summersdale Publishers Ltd

First Skyhorse Publishing paperback edition, 2018

First published in 2011 as *The Cat Lover's Companion*

Copyright © Summersdale Publishers Ltd 2013

Text compiled and written by Lucy York

Title page by Hannah George

Illustrations by Kath Walker

Compilation copyright for additional images © Skyhorse Publishing, Inc 2014

Skyhorse Publishing books may be purchased in bulk at special discounts for sales promotion, corporate gifts, fund-raising, or educational purposes. Special editions can also be created to specifications. For details, contact the Special Sales Department, Skyhorse Publishing, 307 West 36th Street, 11th Floor, New York, NY 10018 or info@skyhorsepublishing.com.

Skyhorse® and Skyhorse Publishing® are registered trademarks of Skyhorse Publishing, Inc.®, a Delaware corporation.

Visit our website at www.skyhorsepublishing.com.

10 9 8 7 6 5 4 3 2 1

Library of Congress Cataloging-in-Publication Data is available on file.

Cover design by Mary Belibasakis
Cover photo: Thinkstock

Paperback ISBN: 978-1-5107-3909-3
Hardcover ISBN: 978-1-62914-778-9
Ebook ISBN: 978-1-62914-898-4

Printed in China

The
CAT LOVER'S
Compendium

Quotes, Facts, and Other
Adorable Purr-ls of Wisdom

Milly Brown

Skyhorse Publishing
A Herman Graf Book

Contents

Introduction

*Prowling his own quiet backyard
or asleep by the fire, he is still only
a whisker away from the wilds.*

Jean Burden

The lives of cats have been intertwined with the lives of humans for thousands of years. For 9,500 years, to be precise: the oldest archaeological evidence of the domestication of the cat dates back to the Neolithic village of Shillourokambos on Cyprus, where the bones of a cat were discovered in the ground next to the remains of a human. The pair's similar state of preservation strongly suggested that they were buried together – perhaps a beloved pet laid to rest with its beloved human.

The fortunes of felines have swung from one extreme to another over the centuries: deified by the Ancient Egyptians, vilified in the Middle Ages, invited into our homes and almost every part of our lives in modern times. Yes, wherever you look around you now, you will find evidence of the cat – in myths and legends, in literary works and art, in popular song and on the silver screen. They are there entertaining our children in cartoons, tempting us in advertising campaigns, adorning our mugs, cushion covers and pyjamas. As pets they are at our elbow as we work, at our table as we eat and cuddling up as we go to sleep. Indeed, for many, a house doesn't quite feel like home unless a cat is present, as Jean Cocteau once so eloquently put it: 'I love cats because I enjoy my home; and little by little, they become its visible soul.'

In this modest tome you will meet a selection of marvellous moggies, some that have narrowly escaped death or been on incredible journeys, others that have made it big in Hollywood, and yet others that have broken world records. You will learn how the Siamese's eyes became crossed and how to predict the weather by watching your cat, and you will read purr-ls of wisdom from feline fans in all walks of life, whether writers like Colette and Ernest Hemingway, artists like Pablo Picasso or celebrities like Drew Barrymore. Cat lovers themselves are a breed apart, and no doubt the tales of the eighteenth-century writer who hand-picked oysters for his prized puss to eat and the millionaires who left it all to their cats will bring a wry smile of recognition to the faces of all you dyed-in-the-wool ailurophiles out there.

So why not settle down in a comfy seat right now with this book and a cup of tea and enjoy some unadulterated cat time? If you're lucky enough, perhaps you'll have a pet puss of your own that will join you and pat at the pages with their paw as you flick through them, warming your knee and adding their soothing purr of approval.

*What greater gift than
the love of a cat?*

Charles Dickens

Our Furry Friends

Our perfect companions never have fewer than four feet.

Colette

How to Befriend a Feline

- Let the cat come to you – if you force your way into its personal space, the cat will feel threatened.

- Avoid making eye contact, keep your distance and adopt a pose to make yourself appear small, for example by crouching down side-on.

- Look out for friendly body language – tail up and curled in greeting and paw raised.

- When the cat feels safe enough to approach, stay crouching down and hold out your hand, still without looking directly at him.

- Let him sniff you for a while. If he is relaxed, test the water with a brief, gentle stroke under the chin – not the top of the head, as cats may find this threatening.

- Build up contact slowly over time – let the cat take the initiative. When stroking, at first stick to safe areas such as the back and sides, avoiding sensitive parts like the head, tummy and legs.

- When you've got to know a cat well, try stroking the base of its tail – most cats love this and will arch their backs to show their pleasure. Scent glands are positioned in this area, and stroking will cause the cat's scent to be released onto you, further strengthening your bond.

- Never pick a cat up until you have gained its trust. When you do, use your arm to support its legs so that it feels secure, and don't hold on too tightly as this will make the cat feel claustrophobic.

- Cats feel insecure when there is tension in the air, or if they hear raised voices, so it's important to maintain a calm environment to make the cat feel at ease.

- Remember that every cat is unique. Some enjoy the company of humans; others do not. Some may be friendly for brief periods, but spend the rest of their day doing other things alone. To build a harmonious relationship, discover what your cat prefers and respect its wishes.

It is no easy matter to win a cat's love, for cats are philosophical, sedate, quiet animals, fond of their own way, liking cleanliness and order, and not apt to bestow their affection hastily. They are quite willing to be friends, if you prove worthy of their friendship, but they decline to be slaves. They are affectionate, but they exercise free will, and will not do for you what they consider to be unreasonable.

Once, however, they have bestowed their friendship, their trust is absolute, and their affection most faithful. They become one's companions in hours of solitude, sadness, and labour. A cat will stay on your knees a whole evening, purring away, happy in your company and careless of that of its own species. In vain do mewings sound on the roofs, inviting it to one of the cat parties where red herring brine takes the place of tea; it is not to be tempted and spends the evening with you. If you put it down, it is back in a jiffy with a kind of cooing that sounds like a gentle reproach. Sometimes, sitting up in front of you, it looks at you so softly, so tenderly, so caressingly, and in so human a way that it is almost terrifying, for it is impossible to believe that there is no mind back of those eyes.

Théophile Gautier, *My Private Menagerie*

Cats at firesides live luxuriously and are the picture of comfort.

Leigh Hunt

*His friendship is not easily won
but it is something worth having.*

Michael Joseph

*If there was any petting to be
done... he chose to do it. Often he
would sit looking at me, and then,
moved by a delicate affection, come
and pull at my coat and sleeve until
he could touch my face with his
nose, and then go away contented.*

Charles Dudley Warner

Stately, kindly, lordly friend,
Condescend
Here to sit by me, and turn
Glorious eyes that smile and burn,
Golden eyes, love's lustrous meed,
On the golden page I read.

All your wondrous wealth of hair,
Dark and fair,
Silken-shaggy, soft and bright
As the clouds and beams of night,
Pays my reverent hand's caress
Back with friendlier gentleness.

Dogs may fawn on all and some
As they come;
You, a friend of loftier mind,
Answer friends alone in kind.
Just your foot upon my hand
Softly bids it understand.

 Algernon Charles Swinburne, from 'To a Cat'

*The cat does not offer services.
The cat offers itself.*

William S. Burroughs

*The key to a successful new
relationship between a cat and
human is patience.*

Susan Easterly

Don't let anyone tell you loving a cat is silly. Love, in any form, is a precious commodity.

Barbara L. Diamond

Rescued a little kitten that was perched in the sill of the round window at the sink over the gasjet and dared not jump down. I heard her mew a piteous long time till I could bear it no longer; but I make a note of it because of her gratitude after I had taken her down, which made her follow me about and at each turn of the stairs as I went down leading her to the kitchen ran back a few steps up and try to get up to lick me through the banisters from the flight above.

From the diary of Gerard Manley Hopkins

From Tabbies to Tortoiseshells

Yes, there they were, big cats, very big cats, middling-sized cats, and small cats, cats of all colours and markings...

Harrison Weir

It's Show Time!

The cat had become a popular domestic companion by the early nineteenth century, and by the late Victorian era people began breeding cats for show. Cat shows, where pedigree felines were admired and given awards for their distinctive qualities and beauty, took off in earnest and were patronised by nobility and even, on occasion, Queen Victoria herself. Harrison Weir, the founding president of the National Cat Club, staged the very first major cat show, held at Crystal Palace in London on 13 July 1871. On the subject he wrote: 'It is many years ago that, when thinking of the large number of cats kept in London alone, I conceived the idea that it would be well to hold "Cat Shows", so that the different breeds, colours, markings, etc., might be more carefully attended to, and the domestic cat, sitting in front of the fire, would then possess a beauty and an attractiveness to its owner unobserved and unknown because uncultivated heretofore.'

Ancestral Tails

- How did the Manx cat lose its tail? Some say that when the Irish invaded the Isle of Man they stole kittens' tails and wore them in their helmets as plumes. Others claim that the Manx was the last animal to board the Ark – it was out hunting mice and didn't come right away when Noah called. The cat leapt aboard just in time but his tail got stuck in the closing door and was cut off.

- In the twelfth century, watered silk was manufactured in the quarter of Baghdad known as 'Attabiy'. The word 'tabby' is thought to originate from this name, because the cat's stripes resemble the moiré pattern of the fabric.

- For a long time Burmese cats were said to cry when their companions died and to be the only felines that could shed tears. However, cats in general do have tear ducts.

- The Egyptian Mau cats that are bred in Great Britain carry marks on their brows that resemble the sacred scarab beetle of Ancient Egypt. Breeders aim to produce cats that resemble as closely as possible those depicted in ancient Egyptian wall paintings.

- A monk was charged with watching over a golden goblet from which Buddha was supposed to have sipped. One day he got tipsy and wandered off, leaving two Siamese temple cats on duty. While one cat set off to find the monk, the other mounted guard, staring so long at the precious goblet that its eyes went crossed. Eventually the little cat could stay awake no longer, and so she curled her tail around the goblet tightly as she slept. When she awoke there was a permanent kink in her tail, and her eyes remained crossed forever.

Papa's Polydactyls

Nobel-prize-winning author Ernest Hemingway was a famous lover of polydactyl cats, which have more than the usual number of toes on their paws due to a genetic mutation. He was given a six-toed cat by a ship's captain, and eventually had quite a collection of them. Because of his love for these animals, they are often known informally as 'Hemingway cats'. When he died in 1961, his former home in Key West, Florida, became a museum and permanent home for his treasured felines.

The Persian Cat is a variety with hair very long, and very silky, perhaps more so than the Cat of Angora; it is however differently coloured, being of a fine uniform grey on the upper part, with the texture of the fur as soft as silk, and the lustre glossy; the colour fades off on the lower parts of the sides, and passes into white, or nearly so, on the belly. This is, probably, one of the most beautiful varieties, and it is said to be exceedingly gentle in its manners.

 **Charles Henry Ross,** *The Book of Cats*

*The Angora cat is most prized.
She is fed with the greatest care,
and, in all respects, is treated like
a respected member of the family;
and noticed, of course, by visitors.*

Eliza Lee Follen

*When Mother Nature saw fit to
remove the tail of the Manx, she
left, in place of the tail, more cat.*

Mary E. Stewart

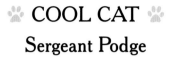 COOL CAT
Sergeant Podge

Norwegian Forest Cat Sergeant Podge had lived with his owner, Liz Bullard, for twelve years. When he disappeared one day from Liz's home in Talbot Woods, Bournemouth, she was worried and began calling her neighbours, asking if any of them had seen him. To Liz's relief, later that day an elderly lady who lived about a mile and a half away rang to say that she had found the missing black cat.

Sergeant Podge was collected from the lady's house and brought home, but a few days later he had gone again. Liz rang the old lady, only to discover that Sergeant Podge was sitting outside her house once again. Since then it became routine for Liz to drop her son off at school, then pick up Sergeant Podge from the same spot on the pavement outside the old lady's house between eight o'clock and quarter past, every morning. When Liz pulled up in her car, she would simply open the door and Sergeant Podge would hop in for the ride home. Of course, being a creature of habit the cat would make his trip every day – no chance of a lie-in for Liz on school holidays and weekends.

It was never discovered precisely why the cat adopted this pattern, though Liz suspected that he might be on the lookout for treats as a woman who used to live nearby had a habit of feeding him sardines. Sergeant Podge became somewhat of a celebrity after tales of his exploits were reported in the national press.

Norwegian Forest Cats have two coats – a thick, woolly undercoat to keep them warm and a waterproof overcoat to keep them dry, making them perfectly adapted for the great outdoors. Though the name of the breed suggests wild origins, it has in fact been a domesticated breed for many centuries. Norwegian Forest Cats enjoy human company and make great pets but, as Sergeant Podge proved, they also love the freedom of exploring the outside world.

What a Beautiful Pussy You Are!

Like a graceful vase, a cat, even when motionless, seems to flow.

George F. Will

Glamour Pusses

Cats don't look perfectly sleek all the time by magic – they spend a significant part of their day grooming themselves from the tip of their nose to the tip of their tail, in order to maintain their cleanliness and appearance. Their rough tongues are perfectly designed to remove loose hairs and dirt, leaving their coats glossy. Owners of long-haired pedigree breeds, however, have to put in a lot of work to help their pampered puss remain tangle free and bushy-tailed. Nowadays, there is the option of taking your cat into a pet grooming centre for a whole range of beauty treatments, from the essentials such as nail clipping and ear cleaning right down to the rather bizarre practice of shaving and fur shaping to give your cat a unique look.

The Owl and the Pussy-cat went to sea
In a beautiful pea-green boat,
They took some honey, and plenty of money,
Wrapped up in a five-pound note.
The Owl looked up to the stars above,
And sang to a small guitar,
'O lovely Pussy! O Pussy, my love,
What a beautiful Pussy you are,
You are
You are!
What a beautiful Pussy you are!'

Edward Lear, from 'The Owl and the Pussycat'

*Two things are aesthetically
perfect in the world – the clock
and the cat.*

Emile-Auguste Chartier

*Which is more beautiful – feline
movement or feline stillness?*

Elizabeth Hamilton

Who's a Pretty Pussy?

If you own a cat you will no doubt consider it to be the most beautiful in the world – however, here's a list of some interesting breeds that are renowned for their unique looks:

The Scottish Fold: these round-faced cuties have a gene that makes their ears fold down which, like the lop-eared rabbit, has done a lot to endear them to potential owners.

Toyger: as the name suggests, these cats resemble pint-sized tigers, having been bred from short-haired tabbies to acquire the distinctive amber and black markings of their jungle-roaming cousins.

Chausie: these sleek-looking felines retain a lot of their jungle cat features and behaviour thanks again to selective breeding – they are, however, just as much at home on the arm of a sofa than on the branch of a tree!

Long may you love your pensioner mouse,
Though one of a tribe that torment the house:
Nor dislike for her cruel sport the cat,
Deadly foe both of mouse and rat;
Remember she follows the law of her kind,
And Instinct is neither wayward nor blind.
Then think of her beautiful gliding form,
Her tread that would scarcely crush a worm,
And her soothing song by the winter fire,
Soft as the dying throb of the lyre.

 **William Wordsworth, 'Loving and Liking'**

Everything a cat is and does physically is to me beautiful, lovely, stimulating, soothing, attractive and an enchantment.

Paul Gallico

A cat is a diagram and pattern of subtle air.

Doris Lessing

In its flawless grace and superior self-sufficiency I have seen a symbol of the perfect beauty and bland impersonality of the universe itself, objectively considered, and in its air of silent mystery there resides for me all the wonder and fascination of the unknown.

H. P. Lovecraft

The Patter of Little Paws

No matter how much cats fight, there always seems to be plenty of kittens.

Abraham Lincoln

A Cuddle a Day

Between the ages of two and seven weeks is the stage in kittens' development when they become socialised and habituated to human contact. So to make sure a kitten grows up to be the perfect friendly, interactive pet, make the most of this optimum time by handling your mini feline for at least 40 minutes daily. Ideally it should have contact with a minimum of four people of different genders and ages, and you should also ensure it encounters a wide range of situations.

The Legend of Pussywillow

An old Polish legend goes that there was once a litter of kittens that was thrown into the river to drown. The sorrowful cries of the mother cat watching helplessly from the bank were heard by willows growing at the river's edge. They dipped their long branches into the waters, and the tiny kittens clung on and were pulled to safety. Ever since, in springtime the pussywillow puts out tiny velvety buds where the kittens had held on for their lives; these are known as catkins.

They say the test of literary power is whether a man can write an inscription. I say 'Can he name a kitten?'

Samuel Butler

Kittens believe that all nature is occupied with their diversion.

Francois-Augustin Paradis de Moncrif

A kitten is so flexible that she is almost double; the hind parts are equivalent to another kitten with which the forepart plays. She does not discover that her tail belongs to her until you tread on it.

Henry David Thoreau

An ordinary kitten will ask more questions than any five-year-old.

Carl Van Vechten

That way look, my infant, lo!
    What a pretty baby-show!
See the kitten on the wall,
Sporting with the leaves that fall,
Withered leaves – one, two, and three –
From the lofty elder tree!
Though the calm and frosty air
Of this morning bright and fair,
Eddying round and round they sink
Softly, slowly: one might think
From the motions that are made,
Every little leaf conveyed
Sylph or faery hither tending,
To this lower world descending,
Each invisible and mute,
In his wavering parachute
But the Kitten, how she starts,
Crouches, stretches, paws, and darts!
First at one, and then its fellow,
Just as light and just as yellow;
There are many now – now one –

Now they stop; and there are none.
What intenseness of desire,
In her upward eye of fire!
With a tiger-leap half-way,
Now she meets the coming prey.
Lets it go as fast, and then;
Has it in her power again.
Now she works with three or four,
Like an Indian conjuror;
Quick as he in feats of art,
Far beyond in joy of heart.
Where her antics played in the eye,
Of a thousand standers-by,
Clapping hands with shout and stare,
What would little Tabby care!
For the plaudits of the crowd?
Over happy to be proud,
Over wealthy in the treasure
Of her exceeding pleasure!

William Wordsworth,
'The Kitten and the Falling Leaves'

A kitten is the most irresistible comedian in the world. Its wide-open eyes gleam with wonder and mirth.

Agnes Repplier

One thing was certain, that the WHITE kitten had had nothing to do with it:– it was the black kitten's fault entirely. For the white kitten had been having its face washed by the old cat for the last quarter of an hour (and bearing it pretty well, considering); so you see that it COULDN'T have had any hand in the mischief.

The way Dinah washed her children's faces was this: first she held the poor thing down by its ear with one paw, and then with the other paw she rubbed its face all over, the wrong way, beginning at the nose: and just now, as I said, she was hard at work on the white kitten, which was lying quite still and trying to purr – no doubt feeling that it was all meant for its good.

But the black kitten had been finished with earlier in the afternoon, and so, while Alice was sitting curled up in a corner of the great arm-chair, half talking to herself and half asleep, the kitten had been having a grand game of romps with the ball of worsted Alice had been trying to wind up, and had been rolling it up and down till it had all come undone again; and there it was, spread over the hearth-rug, all knots and tangles, with the kitten running after its own tail in the middle.

'Oh, you wicked little thing!' cried Alice, catching up the kitten, and giving it a little kiss to make it understand that it was in disgrace.

🐾 **Lewis Carroll, *Through the Looking-Glass***

There is no more intrepid explorer than a kitten.

Jules Champfleury

Confront a child, a puppy, and a kitten with sudden danger; the child will turn instinctively for assistance, the puppy will grovel in abject submission, the kitten will brace its tiny body for a frantic resistance.

Hector Hugh Munro (Saki)

Who's the Boss?

You are my cat and I am your human.

Hilaire Belloc

How to Train your Cat

- If you want a pet that will roll over, sit up and beg and chase its tail on command, you'd be better off getting a dog. That said, with the right approach cats can be trained in certain behaviours which will help make your relationship a more harmonious one.

- Cats rarely perform through praise alone – by associating a command word with a foodie treat you'll get much better results.

- Never strike or shout at your cat to punish it for undesired behaviour, as it does not understand reprimands in the same way humans do; it will only frighten and alienate it from you.

- Remember the three Cs: Clarity (be clear what you are trying to achieve, and how to go about it), Consistency (do the same thing all the time; for example, don't forbid your cat from sleeping on the bed and then allow it to do so as a treat, as this will cause confusion) and Cat-orientated (don't expect your cat to do anything that contradicts its natural species behaviour).

- Target training involves teaching a cat to touch a stick with its paw or nose; once this basic step has been achieved you can progress to more advanced behavioural sequences, such as getting your cat to tackle an agility course. With the right cat and patience this can provide a lot of fun, but never pressurise your cat to perform – it should see the sessions as 'play' and be allowed to opt out when it has had enough.

Petulant Peter

H. G. Wells was devoted to his cat, Mr Peter Wells. This feline was said to have pointedly got up and left the room if any of Wells' guests spoke for too long or too loudly.

Not a few celebrated men have been fond of cats, though only an instance or two can be given here. It is related of Mohammed that once when his cat was asleep on a part of his dress, he cut the part off when he wanted to get up, rather than disturb her. Fine old Samuel Johnson used to go out and buy oysters for his pet cat, thinking that the feelings of his servant might be hurt if sent on such an errand.

 **Charles Dickens, *All Year Round***

A cat will do what it wants when it wants, and there's not a thing you can do about it.

Frank Perkins

Of all God's creatures, there is only one that cannot be made the slave of the lash. That one is the cat.

Mark Twain

🐾 COOL CAT 🐾
Mike

Mike helped guard the gates of the British Museum in London for the twenty years of his life, from February 1909 to January 1929. He has been described as possibly the most famous British cat of the twentieth century.

When Mike was a kitten, he learned from the museum's house cat how to stalk the pigeons that strayed onto museum property. He would bring them to the housekeeper in exchange for a tasty treat, before the housekeeper released the pigeons outside the grounds unharmed.

At the end of Mike's long and successful life, *Time* magazine devoted two articles to him and Sir Wallis Budge wrote an obituary that appeared in the *Evening Standard*. Near the Great Russell Street entrance to the museum a tombstone was erected with the following inscription: 'He assisted in keeping the main gate of the British Museum from February 1909 to January 1929.'

The wording of the inscription gives no hint of Mike's superior and aloof nature – the celebrated cat was known to bestow his attentions very selectively. He would shun female visitors to the museum and had a strong dislike for dogs. He was also very particular about who was allowed to feed him; in his later years, only the official gatekeeper and Sir Ernest A. Wallis Budge were chosen for this privilege.

You can keep a dog; but it
is the cat who keeps people,
because cats find humans useful
domestic animals.

George Mikes

I have noticed that what cats most
appreciate in a human being is...
his or her entertainment value.

Geoffrey Household

The Paw is Mightier Than the Pen

If you want to be a psychological novelist and write about human beings, the best thing you can do is own a pair of cats.

Aldous Huxley

I and Pangur Ban, my cat,
'Tis a like task we are at;
Hunting mice is his delight,
Hunting words I sit all night.

Oftentimes a mouse will stray
Into the hero Pangur's way;
Oftentimes my keen thought set
Takes a meaning in its net.

'Gainst the wall he sets his eye
Full and fierce and sharp and sly;
'Gainst the wall of knowledge I
All my little wisdom try.

When a mouse darts from its den,
O how glad is Pangur then!
O what gladness do I prove
When I solve the doubts I love.

So in peace our tasks we ply,
Pangur Ban, my cat, and I;
In our arts we find our bliss
I have mine and he has his.

Practice every day has made
Pangur perfect in his trade;
I get wisdom day and night,
Turning darkness into light.

Anonymous ninth-century monk, 'Pangur Ban'

Cats are dangerous companions for writers because cat-watching is a near-perfect method of writing avoidance.

Dan Greenburg

He was very fond of books, and when he found one open on the table, he would lie down by it, gaze attentively at the page and turn the leaves with his claws; then he ended by going to sleep, just as if he had really been reading a fashionable novel. As soon as I picked up my pen, he would leap upon the desk, and watch attentively the steel nib scribbling away on the paper, moving his head every time I began a new line. Sometimes he endeavoured to collaborate with me, and would snatch the pen out of my hand, no doubt with the intention of writing in his turn, for he was as æsthetic a cat as Hoffmann's Murr. Indeed, I strongly suspect that he was in the habit of inditing his memoirs, at night, in some gutter or another, by the light of his own phosphorescent eyes. Unfortunately, these lucubrations are lost.

 **Théophile Gautier,** *My Private Menagerie*

A catless writer is almost inconceivable.

Barbara Holland

Poets generally love cats – because poets have no delusions about their own superiority.

Marion Garretty

Because of our willingness to accept cats as superhuman creatures, they are the ideal animals with which to work creatively.

Roni Schotter

Literary Cats

Alexandre Dumas

Mysouff I, Dumas' cat, had an uncanny sense of timing. He would accompany the writer to a certain spot on his way to work and would appear there again to accompany Dumas home, even on days when the writer worked until late.

Charles Dickens

When pet Williamina had kittens, Dickens gave them all to new homes, save one which was deaf. Known as 'The Master's Cat' by the servants, this favoured kitten would snuff out the author's candle when he was at work in order to get his attention.

Colette

The French writer was a celebrated cat lover, and cats feature in many of her novels. On a busy visit to New York, she was returning to her hotel one night when she spied a cat sitting in the street. She went straight over to talk to it, and the two of them mewed to each other for a friendly minute. Colette then turned to her human companion and said with a heartfelt smile: *'Enfin! Quelqu'un qui parle francais!'*

Baudelaire

It is said that whenever the French poet went to a house for the first time, he would be uneasy and restless until he had met the household cat. On seeing it, he would take it up, kiss and stroke it, and be so completely occupied with it he wouldn't reply to anything said to him.

Petrarch

When the writer died in 1374, his son-in-law Francescuolo de Brossano had his beloved pet cat put to death and mummified. It can be seen today in a glass case in Petrarch's study in the Vaucluse in southern France. The niche is decorated with the marble effigy of a cat and an inscription in Latin that says 'Second only to Laura'.

Raymond Chandler

The author of the Philip Marlowe private eye novels, had a black Persian named Taki that he referred to as his 'secretary'. Taki would sit on his manuscripts as he attempted to revise them.

Moggy Myths
and Magic

*I wish I could write as
mysterious as a cat.*

Edgar Allan Poe

We had birds, gold-fish, a fine dog, rabbits, a small monkey, and a cat.

The latter was a remarkably large and beautiful animal, entirely black, and sagacious to an astonishing degree. In speaking of his intelligence, my wife, who at heart was not a little tinctured with superstition, made frequent allusion to the ancient popular notion, which regarded all black cats as witches in disguise. Not that she was ever serious upon this point – and I mention the matter at all for no better reason than that it happens, just now, to be remembered.

Pluto – this was the cat's name – was my favorite pet and playmate. I alone fed him, and he attended me wherever I went about the house. It was even with difficulty that I could prevent him from following me through the streets.

 **Edgar Allan Poe, 'The Black Cat'**

Cat Superstitions From Around the World

America: Dreaming of a white cat brings good luck.

China: A light-haired cat will bring its owner silver, a dark-haired cat will bring its owner gold.

France: If a maiden steps on a cat's tail, it will take her a year longer to find a husband.

Germany: If a sick man sees two cats fighting, death is near.

Holland: If a cat washes behind its ears, you can expect visitors.

Italy: A cat sneezing is a good omen for everyone who hears it.

Russia: To ensure happiness in your new home, a cat must move in with you

Scotland: A strange black cat on your porch brings prosperity.

Wales: Those who feed cats will have sun on their wedding day.

North Africa: An ear of corn is hung behind the door for good fortune – a cat nibbling it will bring extra luck.

Cats, no less liquid than their shadows, offer no angles to the wind.

A. S. J. Tessimond

I cannot deny that a cat lover and his cat have a master/slave relationship. The cat is the master.

Arthur R. Kassin

The cat went here and there
and the moon spun round like a top,
and the nearest kin of the moon,
the creeping cat, looked up.
Black Minnaloushe stared at the moon,
for, wander and wail as he would,
the pure cold light in the sky
troubled his animal blood.

Minnaloushe runs in the grass
lifting his delicate feet.
Do you dance, Minnaloushe, do you dance?
When two close kindred meet,
what better than call a dance?
Maybe the moon may learn,
tired of that courtly fashion,
a new dance turn.

Minnaloushe creeps through the grass
from moonlit place to place,
the sacred moon overhead
has taken a new phase.
Does Minnaloushe know that his pupils
will pass from change to change,
and that from round to crescent,
from crescent to round they range?

Minnaloushe creeps through the grass
alone, important and wise,
and lifts to the changing moon
his changing eyes.

William Butler Yeats, 'The Cat and the Moon'

The cat has always been associated with the moon. Like the moon it comes to life at night, escaping from humanity...

Patricia Dale-Green

He lives in the halflights in secret places, free and alone – this mysterious little great being...

Margaret Benson

Beware the Cat

Witches are said to keep black cats as familiars, and even to be able to transform themselves into feline form. During the reigns of Elizabeth I and James I witch-hunts were common. At that time it simply wasn't safe to own a cat – especially if you were female and lived alone – and many cats were tortured and killed. Two possible associations could have been at play: pagan goddess Freya was said to ride in a chariot drawn by two black cats, and the Romans had linked the Egyptian cat goddess Bast with their own goddess of the moon, Diana. Diana is also linked with Hecate, goddess of witches. At a time when the Church was building its power in Europe, these non-Christian figureheads were demonised. Happily, once the superstitious fervour of these times died down cats once more came to be valued for their rodent-hunting skills and found their way back onto our hearths.

Storm's A-coming

Cats have been frequent fixtures on seafaring vessels throughout the ages, thanks to their rat-catching prowess, but did you know that they also doubled up as weathervanes? Here are just a few of the beliefs that salty sailors held about cats:

- If a cat licked its fur the wrong way, hail was on its way.

- When a cat sneezed you could be sure it was about to rain – and it would come from the direction in which the cat's nose was pointing when it sneezed.

- A hefty gale was coming if the ship's cat was seen frisking on deck.

- Cats were said to be able to raise storms with their tails, so a cat without a tail was seen as a guarantee of safe sailing weather. Manx cats were considered luckiest of all to have aboard. Sailors' wives were sure to keep their cats well fed and happy, to avoid any gale-inducing tail-wagging.

And here's a handy old English folk rhyme to remember next time you're planning a barbeque: If a cat washes her face o'er the ear, 'Tis a sign that the weather will be clear.

Mrs Pipchin had an old black cat, who generally lay coiled upon the centre foot of the fender, purring egotistically, and winking at the fire until the contracted pupils of his eyes were like two notes of admiration. The good old lady might have been – not to record it disrespectfully – a witch, and Paul and the cat her two familiars, as they all sat by the fire together. It would have been quite in keeping with the appearance of the party if they had all sprung up the chimney in a high wind one night, and never been heard of any more.

 Charles Dickens, *Dombey and Son*

A black cat crossing your path signifies that the animal is going somewhere.

Groucho Marx

Cats are glorious creatures who must on no accounts be underestimated... Their eyes are fathomless depths of cat-world mysteries.

Lesley Anne Ivory

Feline and Fancy-free

The cat, which is a solitary beast,
is single minded and goes its
way alone.

H. G. Wells

A Symbol of Freedom

During the French Revolution of 1789, the cat became a symbol of freedom for the French.

The cat represented independence for the Dutch, who used feline forms on their banners during the Eighty Years' War against Spain.

In Ancient Roman times, the goddess of liberty was depicted holding a cup in one hand and a broken sceptre in the other, with a cat lying at her feet.

Cats like doors left open – in case they change their minds.

Rosemary Nisbet

Women and cats will do as they please, and men and dogs should relax and get used to the idea.

Robert A. Heinlein

He will kill mice and he will be kind to Babies when he is in the house, just as long as they do not pull his tail too hard.

But when he has done that, and between times, and when the moon gets up and night comes, he is the Cat that walks by himself, and all places are alike to him. Then he goes out to the Wet Wild Woods or up the Wet Wild Trees or on the Wet Wild Roofs, waving his wild tail and walking by his wild lone.

Rudyard Kipling,
'The Cat That Walked By Himself'

Incredible Journeys

Some cats have taken the feline spirit of adventure to the extreme and have – whether intentionally or unintentionally – ended up crossing continents and even oceans on their travels.

Kiddo

When Walter Wellman made his attempted Atlantic crossing on the airship *America* in 1910, little kitty Kiddo was aboard. Sadly the airship was 475 miles off its destination when the engines failed and the crew had to cut their journey short, meaning that Kiddo was pipped to the post by Whoopsie, said to have been the first feline to fly across the Atlantic aboard the airship R34 in 1919.

Trim

Trim was the first, and may still be the only, cat to circumnavigate Australia with his owner Matthew Flinders aboard the HMS *Investigator* in the early nineteenth century.

Felix

A cat named Felix made a journey of seven weeks by ship and train from the Middle East to Felixstowe in the UK inside a container. He was thought to have survived by licking condensation from the walls of the container.

Emily

When Emily disappeared from her home in Wisconsin, her owners were certainly surprised three weeks later when she was reported to have been found inside a container of paper bales 4,000 miles away in France! The intrepid moggy was flown back to her family in business class by Continental Airlines.

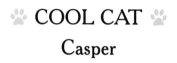

COOL CAT

Casper

A long-haired black and white cat from Devon became famous for his adventures on the buses. Owner Susan Finden, a care worker in her fifties from Plymouth, adopted him from a rescue home in 2002, and had always thought he was quite independent. In fact, she named him after Casper the Friendly Ghost, because he was always disappearing off on his own. But it was when the headstrong feline began his regular excursions on the local First bus service that Susan realised the extent of his spirit of adventure.

Casper had got into the habit of catching the number three service from outside his home and dismounting further along the route. Bus drivers had become so accustomed to the cat that they would automatically stop to let him off at the usual place. No one knows what he got up to when he reached his destination, but he would always turn up there later to catch the bus home.

As Casper was such a well-behaved passenger, bus drivers didn't mind him hopping aboard without a ticket; he would queue patiently with the other passengers, then calmly trot aboard and curl up on his favourite seat. A notice was even put up in the drivers' staffroom, reminding drivers to look after the furry passenger should he board their bus.

This friendly cat, that loved people and seemed fascinated by big vehicles like lorries and buses, met a sad end when he was hit by a car as he crossed the road to board the bus one morning. His owner was devastated at the loss, and he was sorely missed by the bus drivers and his fellow passengers on route number 3.

Meaningful Meows

Meow is like aloha – it can mean anything.

Hank Ketcham

Indeed, the word mete would be more expressively written mieaou. In all these remarks we speak of the vowels as possessing those sounds which are common on the Continent; namely, i like ee, e like ay, a as in father, o as in bone, u as oo, in fool." (English Cyelop., Art. Alphabet.) The reader may try it, and say mi, e, a, o, u, according to the right way of pronouncing. A little practice, with the help of a cat, will soon make the student perfect; but let the student not overdo it, for it is related that a nun in a convent was once seized with a sort of monomaniacal fancy for mewing like a cat, and that in a short time the mania took hold of the other nuns, who went mewing about like cats for some time afterwards.

 **Charles Dickens, from *All the Year Round***

Cat Phrases and Their Meanings

- If cats could understand human speech, they might be surprised to find themselves the subject of a large number of idiomatic sayings.

- If somebody is unusually reticent, you might ask them: 'Cat got your tongue?' This may come from the old Middle Eastern punishment of removing offending body parts – a thief's hand would be cut off and a liar's tongue cut out and the parts would be fed to the cats, as the story goes.

- The Dutch had a phrase that advocated keeping your mouth shut and getting on with the task in hand: 'A cat that meweth catcheth few mice.'

- The saying 'A cat may look at a king' means that an inferior person is not completely restricted in how they might behave before a superior. When that superior is away, the behaviour can even turn to misbehaviour; as they say, 'When the cat's away, the mice will play.'

- The Danish maintain that it takes teamwork and the strength of numbers to defeat a superior adversary: 'It takes a good many mice to kill a cat.'

- If something is the best of the best, it's described as 'the cat's pyjamas'; it has been suggested by some that the phrase came from the name E. B. Katz, an English tailor who was famed for making the best silk pyjamas in the land.

- When you let loose a closely guarded secret people may accuse you of having 'let the cat out of the bag'. Perhaps this comes from medieval times, when at market animals would be sold in sacks. Unscrupulous vendors might trick customers into thinking they were buying a pig or other livestock, when really there was a large cat stowed in the bag. If the customer took a pre-sale peek and revealed the seller's trickery, they would quite literally have let the cat out of the bag.

- Want to really stir things up? Then it's time to 'set the cat among the pigeons' – for example, by revealing a controversial piece of information that will have everyone at each other's throats. This expression may originate from a popular pastime in British India: a wild cat would be put into a pen which contained pigeons and people would place bets on how many birds the cat would dispatch.

'And how do you know that you're mad?'

'To begin with,' said the Cat, 'a dog's not mad. You grant that?'

'I suppose so,' said Alice.

'Well, then,' the Cat went on, 'you see, a dog growls when it's angry, and wags its tail when it's pleased. Now I growl when I'm pleased, and wag my tail when I'm angry. Therefore I'm mad.'

'I call it purring, not growling,' said Alice.

'Call it what you like,' said the Cat. 'Do you play croquet with the Queen to-day?'

'I should like it very much,' said Alice, 'but I haven't been invited yet.'

'You'll see me there,' said the Cat, and vanished.

Alice was not much surprised at this, she was getting so used to queer things happening.

 **Lewis Carroll, *Alice in Wonderland***

Ignorant people think it is the noise which fighting cats make that is so aggravating, but it ain't so; it is the sickening grammar that they use.

Mark Twain

Fur or Against

Cats always know whether people like or dislike them. They do not always care enough to do anything about it.

Winifred Carrière

Famous Cat Lovers

Sir Isaac Newton
Newton was a well-known cat lover and is said to have invented the cat flap, perhaps so his cat could come and go as it pleased without disturbing his work.

Marilyn Monroe
The starlet was devoted to her Persian cat called Mitsou, though at times she found it wasn't easy being a famous cat owner. Of a phone call to the vet she said: 'They think I'm kidding when I say, "This is Marilyn Monroe. My cat is having kittens." They think I'm some kind of nut and hang up.'

Florence Nightingale
owned more than sixty cats over the course of her life and named them all after famous people, such as Bismarck, Disraeli and Gladstone.

Queen Victoria
was very fond of cats and the Victorian era saw a huge rise in their popularity as pets and as the subjects of art and stories.

Sir Winston Churchill

The great Churchill owned several cats, including Nelson, Margate, Tango and a marmalade cat named Jock that ate and slept with him, and attended several wartime cabinet meetings. If Jock was late for dinner, Churchill would send servants to find him and wouldn't start eating until he had arrived.

Record-breaking Affection

On 14 March 2011, Carmen de Aldana of
Guatemala was awarded a Guinness World
Record for the largest cat collection – she
owned 21,321 different cat-related items,
which she had been collecting since 1954.

'I am sorry, very sorry,' said I, 'that you do not like cats. For my part, I think them extremely beautiful, also very graceful in all their actions, and they are quite as domestic in their habits as the dog, if not more so. They are very useful in catching rats and mice; they are not deficient in sense; they will jump up at doors to push up latches with their paws. I have known them knock at a door by the knocker when wanting admittance. They know Sunday from the week-day, and do not go out to wait for the meat barrow on that day; they–' 'Stop,' said my friend, 'I see you do like cats, and I do not, so let the matter drop.' 'No,' said I, 'not so. That is why I instituted this Cat Show; I wish every one to see how beautiful a well-cared-for cat is, and how docile, gentle, and – may I use the term? – cossetty.'

Harrison Weir, *Our Cats and All About Them*

Cat people are different to the extent that they generally are not conformists.

Louis J. Camuti

Cat lovers can readily be identified. Their clothes always look old and well used.

Eric Gurney, *How to Live with a Calculating Cat*

People meeting for the first time suddenly relax if they find they both have cats. And plunge into anecdote.

Charlotte Gray

Famous Cat Haters

An ailurophobe is someone who has an irrational and intense fear of cats, often because they had a traumatic experience involving a cat when they were very young. Famous ailurophobes include Adolf Hitler, Benito Mussolini and Julius Caesar. Other people, however, just seem to have it in for the poor creatures for no apparent reason:

Alexander the Great
The celebrated war-monger who conquered vast areas of the ancient world, is said to have swooned at the sight of a cat.

Johannes Brahms
The composer is reported to have gleefully shot at neighbourhood cats with a bow and arrow from the window of his house in Vienna.

Napoleon Bonaparte
Napoleon was once seen flailing about wildly with his sword and clammy with fear, all because a mere kitten was hiding behind a tapestry in the same room as him.

Ivan the Terrible

Ivan lived up to his name even as a child, when he would amuse himself by hurling cats out of high windows.

Rockwell Sayre

Chicago banker Rockwell Sayre thought cats were 'filthy and useless' and took it upon himself to see that all cats in the world were destroyed. In the 1920s he offered rewards to people for killing cats. Thankfully he died long before coming close to achieving his aim!

Henri III of France

He might have been bold as a lion, but the sight of a humble kitty was enough to make him faint. During his reign 30,000 cats were put to death.

There is, incidentally, no way of talking about cats that enables one to come off as a sane person.

Dan Greenburg

People with insufficient personalities are fond of cats. These people adore being ignored.

Henry Morgan

Eternal Love

Some devoted cat lovers have gone to great lengths to ensure their beloved pet's welfare, even after their own passing:

- In AD 1280, El Daher Beybars, the then sultan of Egypt and Syria, left a garden to be a haven for homeless cats. It was called Gheyt al Qottah, 'The Cats' Orchard'. It eventually fell into ruin, but for centuries afterwards people of the area maintained the tradition of feeding stray cats.

- In the eighteenth century, the Earl of Chesterfield and the Second Duke of Montagu both left their pet cats considerable pensions in their wills.

- A Mrs Walker from England left £3 million to an animal charity, provided the organisation care for her cat for the rest of his life.

- Madame Dupuy, a famous seventeenth-century harpist, made provision for her two cats in her will, and included some very specific instructions about mealtimes: 'They are to be served daily, in a clean and proper manner, with two meals of meat soup, the same as we eat ourselves, but it is to be given them separately in two soup plates. The bread is not to be cut up into the soup, but must be broken into squares about the size of a nut, otherwise they will refuse to eat it. A ration of meat, finely minced, is to be added to it; the whole is then to be mildly seasoned, put into a clean pan, covered close, and carefully simmered before it is dished up.'

- Woodbury Rand, an attorney from Boston, left $40,000 to his cat, Buster, stating that his relatives didn't deserve a penny because of their cruelty to his cat.

- The well-known veterinarian, Louis J. Camuti, reported the case of an elderly widow who wished to be buried with her cat. When the cat died it was placed in a casket and sent to a pet cemetery until such time as she died. However, she subsequently discovered that under state health laws she could not have the cat buried with her in the coffin – and so she asked Dr Camuti to cremate the cat. The ashes were then sewn into the hem of her wedding dress, which she would be buried in.

People that don't like cats haven't met the right one yet.

Deborah A. Edwards

The Artist's Mews

*The smallest feline is
a masterpiece.*

Leonardo da Vinci

Gottfried Mind, the celebrated Swiss painter, was called the 'Cat Raphael', from the excellence with which he painted that animal. This peculiar talent was discovered and awakened by chance. At the time when Frendenberger painted his picture of the 'Peasant Clearing Wood', before his cottage, with his wife sitting by, and feeding her child out of a basin, round which a Cat is prowling, Mind, his new pupil, stared very hard at the sketch of this last figure, and Frendenberger asked with a smile whether he thought he could draw a better. Mind offered to show what he could do, and did draw a Cat, which Frendenberger liked so much that he asked his pupil to elaborate the sketch, and the master copied the scholar's work, for it is Mind's Cat that is engraved in Frendenberger's plate. Prints of Mind's Cats are now common.

 **Charles Henry Ross, *The Book of Cats***

Cats and Canvas

French artist Henri Matisse loved the company of cats. Whenever he was ill and had to stay in bed, his favourite black cat would keep him company there.

The Swiss artist Theophile Steinlen had a home in Paris known as 'Cats Corner'. Cats were the subject of many of his paintings, drawings and sculptures.

Pierre Auguste Renoir was another French artist who admired the feline race, and he portrayed them in several paintings.

Black and white cat Peter belonged to the wife of English artist Louis Wain. Wain taught Peter tricks to amuse his wife, and the cat appeared in many of his early drawings. The artist later acknowledged the cat's influence on his work: 'To him properly belongs the foundation of my career, the developments of my initial efforts, and the establishing of my work.'

Muses for Musicians

- *L'enfant et les sortilèges* is an opera written by French classical composer Maurice Ravel with the French writer Colette. It includes a bravura cat duet sung by the characters Tom Cat and She Cat in a feline-sounding language.

- Domenico Scarlatti's cat Pulcinella often scampered across his harpsichord and gave the composer the opening notes for 'Fugue in G Minor, L499', or the 'Cat's Fugue'.

- 'Delilah', released on Queen's album *Innuendo*, was written by Freddie Mercury about his favourite tortoiseshell cat.

- Andrew Lloyd Webber, composer of the musical *Cats*, was writing the sequel to *Phantom of the Opera* in 2007 when his six-month-old Turkish Van, Otto, wiped out the score with the stroke of a paw. Said Webber: 'I was trying to write some new music, he got into the grand piano, jumped onto the computer and destroyed the entire score for the new "Phantom" in one fell swoop.'

- Cats have also featured in numerous popular songs, including 'The Siamese Cat Song' by Peggy Lee, 'Walking My Cat Named Dog' by Norma Tanega, 'Tom Cat' by The Rooftop Singers, 'Smelly Cat', sung by the character Phoebe Buffay in the sitcom *Friends*, and the traditional song 'The Cat Came Back'.

- The Guinness World Record for the earliest piano concerto for a cat was awarded to 'Catcerto' by Mindaugas Piecaitis of Lithuania. This four-minute piano piece for chamber orchestra debuted at the Klaipeda Concert Hall in Lithuania on 5 June 2009. The orchestra accompanied a video recording of Nora the cat pawing at notes on a piano.

- One day when Frédéric Chopin's cat walked across his keyboard the composer liked the impromptu melody so much that he composed a whole piece around it, the 'Cat Waltz'.

COOL CAT
Nora

Betsy Alexander, a piano instructor from Philadelphia, adopted her grey tabby cat from a rescue centre in Cherry Hill, New Jersey. She named her Nora, after the artist Leonora Carrington, and the cat soon proved to have her own creative tendencies.

Betsy would spend a lot of time in her studio teaching, and the cat showed an interest in the piano straight away. Sometimes she would dance in circles on top of the piano while Betsy played. Then, when she was one year old, she climbed up on the bench in front of a Yamaha Disklavier piano and began to press the keys with her paws. From that point on she would regularly 'play' the piano side by side with Betsy, who has two of the instruments lined up in her studio.

In 2007 Betsy made a film of Nora playing the piano and posted it on the website YouTube. It received 17 million hits and attracted the attention of the media – soon the clip was being shown on VH1, Conan O'Brien, Tyra Banks, Ellen, Martha Stewart, and many more. Nora even appeared on *The Today Show* playing the piano in a live video feed, to prove that the YouTube clip was not faked.

Betsy said that she thought Nora enjoyed the attention she got from playing the piano, but that really she did it to please herself, as she would play when she was alone in the room.

Nora, dubbed 'The Piano Cat', received well-wishes from the Piano Man himself, Billy Joel. There is a website dedicated to her, where she has her own blog. She also has her own DVD, a book called *Nora the Piano Cat's Guide to Becoming a Good Musician* has been published, and a CD has been released that features a track that incorporates her playing.

Whether they be the musician cats in my band or the real cats of the world, they all got style.

Ray Charles

Kitty IQ

I've met many thinkers and many cats, but the wisdom of cats is infinitely superior.

Hippolyte Taine

Who's a Clever Pussycat?

Felines aren't usually thought of as receptive to being taught tricks, but in fact they are better at it than dogs because they are able to learn by observing very carefully. For example, many cats learn how to jump at a door and open it by pulling down the handle, just by watching humans. They are very quick to notice which of their actions attract their human's attention; usually this is used to manipulate their human – say, by scratching at furniture when they want to go out. This capacity to learn can be transformed into astounding tricks by a patient trainer who is willing to let the cat learn on its own terms:

- Signor Cappelli's performing musical cats were billed as 'the greatest wonder in England' back in 1829. These talented felines wowed audiences with their trapeze and juggling skills, along with more practical talents such as grinding rice, roasting coffee and turning a spit to draw water out of a well.

- Yuri Kouklachev of the Moscow State Circus trained a troupe of cats to jump over obstacles, do handstands and even play chess. He found success by training the animals at night, when he claimed they were more receptive because of their nocturnal nature.

*Cats are a mysterious kind of folk.
There is more passing in their
minds than we are aware of.*

Sir Walter Scott

*The smart cat doesn't let on
that he is.*

H. G. Frommer

As the learned and ingenious Montaigne says like himself freely, When my cat and I entertain each other with mutual apish tricks, as playing with a garter, who knows but that I make my cat more sport than she makes me? Shall I conclude her to be simple, that has her time to begin or refuse to play as freely as I myself have? Nay, who knows but that it is a defect of my not understanding her language (for doubtless cats talk and reason with one another) that we agree no better? And who knows but that she pities me for being no wiser than to play with her, and laughs and censures my folly for making sport for her, when we two play together?

Izaak Walton, *The Compleat Angler*

Wily Cats

From Puss in Boots to the ThunderCats, cats have often been represented as quick-witted characters with an uncanny propensity for coming out on top whatever the situation.

Top Cat
The lead character of the eponymously titled Hanna-Barbera cartoon series had a knack for coming up with outrageous get-rich-quick schemes and was always one step ahead of arch-enemy Officer Dibble.

Macavity
The master criminal cat who features in T. S. Eliot's *Old Possum's Book of Practical Cats* committed a multitude of crimes but never left behind any recriminating evidence and was always miles away from the scene by the time the police showed up.

Salem
Sabrina's feline companion in the TV sitcom *Sabrina, The Teenage Witch* was a black cat who was actually a warlock sentenced to take the form of a talking cat for a period of time. Though the magic-related advice he gave Sabrina often seemed to exacerbate problems for the budding witch, his witty one-liners and sharp retorts always raised a laugh from the audience.

Felix

The Chaplinesque Felix was the first feline star of the silent movie screen. The quick and resourceful cat would always triumph, and is arguably the most famous cartoon cat of all time.

Crookshanks

In J. K. Rowling's *Harry Potter* series of books, Hermione Granger's pet cat Crookshanks may not be the most attractive puss with his squashed face, but the intelligent creature is able to spot when things aren't what they seem. He is the first to notice something strange about Ron Weasley's pet rat Scabbers, who later turns out to be the duplicitous character Peter Pettigrew in rat-form.

The Cat That Walked by Himself

One of Rudyard Kipling's *Just So Stories*, 'The Cat That Walked by Himself' is set in the time when man lived in caves and had only just begun to domesticate animals. It features a cat that secures a place next to man's hearth by tricking the woman; rather than becoming a servant of men as the dog, horse and cow do, the cat retains the right to still come and go as he pleases.

Feel the Magic, Hear the Roar...

... ThunderCats are loose! Anyone familiar with the eighties animated TV show may remember the joy of those opening lines, accompanied by an action-packed sequence showing the cats doing their stuff, leaping about and wielding their various weapons. Like many TV shows of the time, *ThunderCats* was closely linked with an extensive range of merchandise – everything from action figures to replicas of the lead character Lion-O's sword – all of which made your regular household pet seem a little average!

Monsieur Puss came at last to a stately castle, the master of which was an Ogre, the richest that had ever been known; for all the lands which the King had then gone over belonged to this castle. The Cat, who had taken care to inform himself who this Ogre was, and what he could do, asked to speak with him, saying, he could not pass so near his castle, without having the honour of paying his respects to him.

The Ogre received him as civilly as an Ogre could do, and made him sit down.

'I have been assured,' said the Cat, 'that you have the gift of being able to change yourself into all sorts of creatures you have a mind to; you can, for example, transform yourself into a lion, or elephant, and the like.'

'This is true,' answered the Ogre very briskly, 'and to convince you, you shall see me now become a lion.'

Puss was so sadly terrified at the sight of a lion so near him, that he immediately got into the gutter, not without abundance of trouble and danger, because of his boots, which were ill-suited for walking upon the tiles. A little while after, when Puss saw that the Ogre had resumed his natural form, he came down, and owned he had been very much frightened.

'I have been moreover informed,' said the Cat, 'but I know not how to believe it, that you have also the power to take on you the shape of the smallest animals; for example, to change yourself into a rat or a mouse; but I must own to you, I take this to be impossible.'

'Impossible?' cried the Ogre, 'you shall see that presently,' and at the same time changed into a mouse, and began to run about the floor.

Puss no sooner perceived this, but he fell upon him, and ate him up.

Charles Perrault, 'Puss in Boots',
The Fairy Tales of Charles Perrault

It was not I who was teaching my cat to gather rosebuds, but she who was teaching me.

Irving Townsend

If you would know what a cat is thinking about, you must hold its paw in your hand for a long time.

Jules Champfleury

Dinner is Served

Dogs eat. Cats dine.

Ann Taylor

A cat in distress,
Nothing more, nor less;
Good folks, I must faithfully tell ye,
As I am a sinner,
It waits for some dinner
To stuff out its own little belly.

Percy Bysshe Shelley, from 'Verses On A Cat'

Feline Food Habits

- A cat named Arthur became famous for his distinctive approach to dining. The large white short-haired cat would scoop cat food straight from the tin and then lick it off his paw, and shot to stardom when he featured in Kattomeat's successful TV commercial.

- Becky Page from Tasburgh, Norfolk, was surprised when a black and white stray kitten she adopted turned his nose up at a bowl of fresh chicken. He tucked into some leftover vegetables instead and maintained a diet of vegetarianism ever since. Dante's choice of diet is extraordinary because cats are natural carnivores; there are certain vital nutrients that they can only get in sufficient quantities from meat, such as the essential amino acid taurine and arachidonic acid. Becky had to resort to sneaking bits of meat into Dante's vegetarian meals to ensure he remained a healthy kitty.

- Guinness World Records' heaviest cat was Himmy from Queensland, Australia, who weighed in at a whopping 46.8 pounds. Guinness no longer accepts nominations for this record, because they don't wish to encourage pet owners to overfeed their pets. If the record was still up for grabs, the most likely contender for the title would be Russian cat Katy, reported to weigh around 45 pounds.

There is a propensity belonging to common house-cats that is very remarkable; I mean their violent fondness for fish, which appears to be their most favourite food: and yet nature in this instance seems to have planted in them an appetite that, unassisted, they know not how to gratify: for of all quadrupeds cats are the least disposed towards water; and will not, when they can avoid it, deign to wet a foot, much less to plunge into that element.

Gilbert White, *The Natural History of Selborne*

Egbert seized the milk-jug and poured some of its contents into Don Tarquinio's saucer; as the saucer was already full to the brim an unsightly overflow was the result. Don Tarquinio looked on with a surprised interest that evanesced into elaborate unconsciousness when he was appealed to by Egbert to come and drink up some of the spilt matter. Don Tarquinio was prepared to play many roles in life, but a vacuum carpet-cleaner was not one of them.

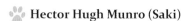 **Hector Hugh Munro (Saki)**

A cat... would check to see if you brought anything to eat, and if not, would turn and walk away, tail held high.

Mike Deupree

Even overweight cats instinctively know the cardinal rule: when fat, arrange yourself in slim poses.

John Weitz

Cat Salad

Cats like the odd bit of greenery to chew on – you might often spy them nibbling on grass in the garden. Keep a pot of catnip or sedum for your cats to munch, but make sure they stay clear of the following plants, which are poisonous (and some potentially fatal) to cats: andromeda, azalea, bittersweet, boxwood, crown of thorns, daphne, dumb cane, elephant ear, English ivy, foxglove, holly, hydrangea, Jerusalem cherry, lantana, laurel, lily-of-the-valley, mistletoe, monkshood, oleander, philodendron, pine needles, poinsettia, privet, rhododendron, snow-on-the-mountain, sheep laurel.

Wonder Cats

There are no ordinary cats.

Colette

Special Senses

Smell
It is thought that cats are able to detect certain medical problems in humans. Some scientists have suggested that up to an hour and a half before a person has a seizure, abnormal electrical activity begins in the part of the brain that regulates heartbeat and perspiration – cats may be able to pick up on these resulting changes. Medical studies have also shown that patterns of biochemical markers have been found in the exhaled breath of cancer patients. With their acute sense of smell, cats would be able to detect these subtle changes and markers, and may alter their behaviour towards the person as a result.

Hearing
Cats have an acute sense of hearing. They can detect a very wide range of frequencies and are able to hear higher pitched sounds than humans and even dogs can hear: anything from 55 Hz up to 79 kHz, a range of 10.5 octaves. This allows cats to hear the ultrasonic calls that many species of rodents make. Their large movable outer ears help to amplify sounds and enable them to sense the direction from which a noise is coming.

Touch
A cat's whiskers (or vibrissae) grow from a special type of follicle connected to nerve endings. These give cats an extra level of perception, allowing them to sense

close objects when visibility is impaired and to navigate narrow gaps safely.

Seismologists have noted that about ten days before an earthquake happens, cats begin to leave the area. This could be because changes in the earth's gravity move the magnetite in their bodies, allowing them to sense the oncoming danger and flee to safety.

Why do Cats Have Nine Lives?

Cats do seem to have an uncanny ability to cheat death and escape from precarious situations, in particular to survive falls from great heights. In Medieval Europe cats were sometimes thrown from high towers as part of rituals – the fact that the cats often walked away seemingly unscathed much amazed spectators and further contributed to the nine lives myth. We now know that cats can right themselves mid air and thereby absorb the shock of a fall by landing on their feet; however, they don't always escape without injury, often sustaining cuts, bruises and fractures if they do fall from very high up.

But why specifically nine lives? The Ancient Egyptians believed in three groups of gods, each made up of nine gods; as the Egyptians cherished cats, the association between felines and the number nine may date from this period. The Norse goddess Freya is also linked with cats and with the number nine – she was said to hold power over the nine worlds of Norse mythology.

Always turn and look when your cat gazes behind you with that intent look in her eyes. Some day there might actually be something there.

Anonymous

It always gives me a shiver when I see a cat seeing what I can't see.

Eleanor Farjeon

Very strange is the power which cats may show of finding their way home by routes which they have never before traversed. We cannot explain this (as it has been sought to explain the like power in dogs), by the power of smell being the predominate sense, so that a passed succession of smells can be retraversed in reverse order, as a number of places seen in succession on a journey may be retraversed in reverse order by ourselves. On the whole, it seems probable that the power in question may be due to a highly developed 'sense of direction', like that which enables some men so much to excel others in finding their way about cities, or that which enables the inhabitants of Siberia to find their way through woods or over hummocky ice, and who, though constantly changing the direction they immediately pursue, yet keep their main direction unchanged.

🐾 **St George Mivart,** *The Cat:*
An Introduction to the Study of Backboned Animals

It has been the providence of Nature to give the cat nine lives instead of one.

Bidpai

Missed By a Whisker

Judging by the scrapes they get into, some cats seem to take the saying that they have nine lives quite literally, as these impressive tales of survival show:

- Three-year-old Lucky survived a twenty-six-storey fall out of his owner's apartment window in Manhattan when a barbeque on a lower storey balcony broke his fall. In fact, 90 per cent of cats that fall from high-rise buildings are reported to survive, though many are left with broken bones and bruises.

- Felix, whose owner's house backed onto farmland in Brentwood, Essex, survived the terrifying ordeal of being pulled through a combine harvester. The poor kitty had to have his tail and right back leg amputated and also contracted the feline form of tetanus, a very rare illness indeed. He made a miraculous recovery and quickly adapted to life on three legs.

- A seafaring cat named Oscar, also later known as Unsinkable Sam, survived the sinking of the German battleship *Bismarck* in 1941. He was rescued by sailors aboard the British destroyer *Cossack*, only to make another narrow escape when the Cossack itself went down later that year. He finally cheated death one more time aboard the *Ark Royal* when that ship was hit by a torpedo and sunk just off Gibraltar, where Oscar lived out the rest of his days. A portrait of this fearless feline can be found at the National Maritime Museum in Greenwich, entitled *Oscar, the Bismarck's Cat*.

🐾 COOL CAT 🐾
Oscar

Oscar, a fluffy, grey and white brindled cat, was adopted by the dementia unit of the Steere House Nursing and Rehabilitation Centre in Providence, Rhode Island, in July 2005.

Oscar settled in quickly and soon began making his own rounds of the home. Staff noticed that the little cat was not generally friendly – in fact, he was quite selective. Sometimes he would just sniff the patients and turn away, and the patients he did curl up next to on the bed would pass away within a few hours. Was he somehow able to predict when patients were about to die?

When he made his thirteenth correct prediction, staff decided to watch his behaviour around a patient who had deteriorated and was exhibiting many signs of imminent death – she had stopped eating, had difficulty breathing and there was a bluish tinge to her legs. However, Oscar didn't seem to be concerned at that point. Ten hours later, Oscar finally appeared at the woman's bedside, and she died within two hours. It seemed that Oscar's skill was much more precise than the medical staff's.

By the age of two, Oscar had predicted the deaths of twenty-five residents. Staff began alerting the families of patients he curled up next to, so that they would have time to visit their relative before the end. Many families seemed thankful for his presence and the early warning and companionship that the cat provided for their dying loved ones.

The most likely explanation for Oscar's ability is that he has learnt to recognise certain scents created by chemical and hormonal changes in the body of a dying patient that are imperceptible to humans. His reasons for staying at a patient's side once he has predicted they are about to die cannot be explained, however; perhaps he feels an urge to comfort them as they leave this world.

A wall plaque was put up in the home in recognition of Oscar's 'compassionate hospice care'. His story has been widely reported in the media, and the publicity has resulted in a surge of letters and emails to Steere House from people claiming to know of other cats with similar abilities.

On the Prowl

Cats too, with what silent stealthiness, with what light steps do they creep towards a bird!

Pliny, *Natural History*

A cat brings you gifts: half a lizard, an eviscerated squirrel, but she means well.

Leonore Fleischer

The cat, with eyes of burning coal, Now Couches 'fore the mouse's hole.

William Shakespeare, *Pericles, Prince of Tyre*

A Cat at Play

Why do cats 'play' with their catches? There are several possible reasons for this. It may be in part because the cat is an inexperienced hunter, and therefore a poor killer; the cat instinctively knows how to catch prey but doesn't know how to finish it off. It could also be a means of weakening its prey, until the cat feels sure it can move in to dispatch his victim without risking a scratch or nip to the face from the frightened target. Many cats bring half-dead prey home for their human owners. A cat views its owner as a littermate, and it presents the prey to be used for practising hunting skills, just as a mother cat would bring home live prey in order to teach her kittens hunting skills.

She sights a Bird – she chuckles –
She flattens – then she crawls –
She runs without the look of feet –
Her eyes increase to Balls –

Her Jaws stir – twitching – hungry –
Her Teeth can hardly stand –
She leaps, but Robin leaped the first –
Ah, Pussy, of the Sand,

The Hopes so juicy ripening –
You almost bathed your Tongue –
When Bliss disclosed a hundred Toes –
And fled with every one.

 **Emily Dickinson, 'Cat'**

*The cat does not negotiate
with the mouse.*

Robert K. Massie

*The clever cat eats cheese and
breathes down rat holes with
baited breath.*

W. C. Fields

My friend Captain Noble, of Maresfield, informs me that he has himself known a cat which was in the habit of catching starlings by getting on a cow's back and waiting till the cow happened to approach the birds, which little suspected what the approaching inoffensive beast bore crouching upon it. He assures me he has himself witnessed this elaborate trick, by means of which the cat managed to catch starlings which otherwise it could never have got near.

St George Mivart, *The Cat:*
An Introduction to the Study of Backboned Animals

Calling Pest Control

Thanks to their excellent hunting skills, many cats have made quite a career as rodent population controllers over the years:

- Minnie was the resident rat-catcher at a sports stadium in London, and caught an average of 2,000 rats a year over the six years she worked there.

- During World War Two, the British Ministry of Supply put out a request for members of the public to volunteer their pet cats for service. Hundreds of cats were sent in by patriotic owners and were put to the task of keeping army stores vermin free.

- Towser was a tabby that worked as mouse-catcher in a Scotch distillery's grain stores and is said to have caught over 23,000 mice in her twenty-three-year-long career there.

A Lesson in Comfort

Cats are connoisseurs of comfort.

James Herriot

There are few things in life more heartwarming than to be welcomed by a cat.

Tay Hohoff

For me, one of the pleasures of cats' company is their devotion to bodily comfort.

Sir Compton Mackenzie

A Calming Influence

Many doctors and medical specialists believe that spending time with cats can help regulate a person's blood pressure and heart rate, and also relieve stress. In the UK and the US there are now organisations that train 'therapy' cats specifically to provide support and comfort to the unwell. These cats are taken by registered volunteers to visit patients in hospitals and care homes. Regular visits from a friendly therapy cat could help a patient who has become withdrawn to come out of themselves, or could improve the memory recall function of elderly patients by triggering memories of pets they've had.

A cat pours his body on the floor like water. It is restful just to see him.

William Lyon Phelps

The ideal of calm exists in a sitting cat.

Jules Reynard

🐾 COOL CAT 🐾
Faith

During World War Two, when Britain was under attack and hundreds of people had died in air raids, or had lost their homes, the story of a brave little cat in the capital brought comfort and hope to many across the land.

When the little stray tabby arrived at St Augustine's Church in London, the rector, Henry Ross, couldn't resist taking her in. He decided to keep her as the church's cat, and named her Faith. She soon became popular with the parishioners, and was often seen lying stretched out at Ross's feet or on the front pew during his sermons.

In August 1940, Faith gave birth to a black and white tom kitten, named Panda. Mother and kitten settled into a basket in the rector's living quarters and seemed to be getting on very well. But one day, Faith began to investigate the different rooms of the house and appeared very restless. Then she moved the kitten down to the basement. When Ross found them there he took them back upstairs where it was warm. But the next day he found them in the basement again. This happened three times, until Ross gave in and moved the basket down there, where the pair happily settled in between stacks of music sheets.

On 9 September, Ross went to Westminster on business. As he journeyed home that evening, the air raid warning sounded and he had to spend the night in a shelter. London was badly bombed that night, and many buildings were destroyed, among them eight churches – including St Augustine's. Only its tower was still standing – the rest had been reduced to a mass of smouldering rubble.

Firemen warned Ross to stay back, but he approached anyway, hoping to find his beloved cats. Then, a faint meowing sound came from within the smoking mound. Ross struggled to move rubble and timbers aside, eventually revealing two filthy, bedraggled, frightened but completely unharmed cats.

Faith's miraculous story was reported in the national press, and tributes came flooding in. The PDSA's Dickin Medal is an honour reserved for military animals, so Faith wasn't eligible to receive it, but Maria Dickin, the award's founder, decided to honour Faith with a special silver medal in recognition of her steadfast courage. She was the first cat to receive such an accolade for bravery.

The weather was cold, and the sick lady had the dreadful chills, that accompany the hectic fever of consumption. She lay on the straw-bed, wrapped in her husband's great coat, with a large tortoiseshell cat in her bosom. The wonderful cat seemed conscious of her great usefulness. The coat and the cat were the sufferer's only means of warmth; except as her husband held her hands, and her mother her feet.

An account of the death of Edgar Allan Poe's wife Virginia Clemm in 1844, written by a Mrs Grove-Nichols

Top Cats

Thou art indeed... the Great Cat.

Part of the inscription on the
Royal Tombs at Thebes

Class she certainly was, from her tapered black head, beautiful as an Egyptian queen carved out of ebony, to the tip of her elegant whip tail. I thought she was the loveliest animal I had ever seen, and when the old man went on to tell us how she climbed the curtains like a monkey when the fit took her, perched on the rail, and refused to come down, or went round the room leaping from the top of the piano to the mantelpiece like a racehorse, I knew I was lost.

Doreen Tovey, *Cats in the Belfry*

Dogs believe they are human.
Cats believe they are God.

Jeff Valdez

If man could be crossed with the
cat it would improve the man, but
it would deteriorate the cat.

Mark Twain

All Hail, Great Cat

- The Romans believed that their god Venus transformed a cat into a beautiful woman and named her Ailuros (meaning cat in Greek). Ailuros challenged Venus's beauty and Venus, furious, turned her back into a cat.

- In theory, the Hindu religion requires the housing and feeding of one cat by all the faithful.

- In Ancient Egypt, the goddess Bast (also know as Pasht, or Bastet) was depicted with the head of a cat. Sacred cats resided at her temple in the holy city of Bubastis, and at that time all cats were protected. Bast was also the goddess of fire, and because a dry cat's fur gives off sparks of electricity when stroked, the cat was associated with fire through her.

- Egyptians also linked the cat with the sun god Ra – they believed that cats' eyes captured the fire of the sun during the day and reflected it back at night. This signified that the sun would return in the morning. The solar and lunar gods would be enraged if a cat was killed, threatening the end of the world and eternal darkness. Killing a cat was hence punishable by death.

Over the hearth with my 'minishing eyes I
muse; until after
the last coal dies.
Every tunnel of the mouse,
every channel of the cricket,
I have smelt,
I have felt
the secret shifting of the mouldered rafter,
and heard
every bird in the thicket.
I see
you
Nightingale up in the tree!
I, born of a race of strange things,
of deserts, great temples, great kings,
in the hot sands where the nightingale never sings!

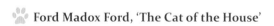 **Ford Madox Ford, 'The Cat of the House'**

Stars of the Big Screen

The Patsy Awards were run from 1951 to 1986 by the American Humane Association and were the equivalent of the Oscars for animal stars. The award, which celebrated the best animal performance in a film, was given to felines five times over the years:

1952: *Rhubarb* – A red tabby named Orangey starred in the role of Rhubarb, a cat who is the sole beneficiary of a millionaire and thereby inherits a baseball team.

1959: *Bell, Book and Candle* – This romantic comedy starring Kim Novak and James Stewart, about a woman named Gillian who casts a love spell, featured a Siamese playing the role of Pyewacket, Gillian's familiar.

1962: *Breakfast at Tiffany's* – Orangey was up for his second gong after playing the ubiquitous 'Cat' in this Audrey Hepburn classic.

1966: *That Darn Cat* – Siamese cat Syn bagged the award for his performance in this film, but the character DC, a cat who becomes key to snaring a band of robbers, was in fact played by two Siamese cats. Syn also starred as Tao in the 1963 film *The Incredible Journey*.

1975: *Harry and Tonto* – Elderly widower Harry Coombes is forced out of his apartment when his building is condemned and embarks on a journey across America with his pet cat Tonto in this touching road trip movie.

In the beginning, God created man, but seeing him so feeble, He gave him the cat.

Warren Eckstein

The reason cats climb is so that they can look down on almost every other animal...

K. C. Buffington

Dear creature by the fire a-purr,
Strange idol, eminently bland,
Miraculous puss! As o'er your fur
I trail a negligible hand,
And gaze into your gazing eyes,
And wonder in a demi-dream,
What mystery it is that lies,
Behind those slits that glare and gleam,
An exquisite enchantment falls
About the portals of my sense;
Meandering through enormous halls,
I breathe luxurious frankincense...

 Lytton Strachey, from 'The Cat'

Cats were put into the world to disprove the dogma that all things were created to serve man.

Paul Gray

Pitch Purr-fect

To err is human, to purr is feline.

Robert Byrne

What's in a Purr?

Exactly how cats produce that distinctive feline sound known as purring remains a scientific mystery, and neither has the purpose of the purr been definitively determined. As kittens, the combination of purring and paddling the mother's belly with their tiny front paws is thought to induce the milk let-down reflex and enable them to suckle. A grown-up cat may purr when relaxed and contented, and they often do so in response to being stroked by a human they trust, signalling their receptiveness to social contact. However, cats also purr when they are extremely distressed, for example if they are ill, injured or anxious – this explains why some cats purr frantically when being examined by a vet. Yet another explanation may be hidden in the enigmatic purr: studies of the healing properties of low-frequency vibrations have shown that a cat's purr vibrates at a frequency considered beneficial for bone growth, fracture healing and even pain relief.

Making Purr-fect Sense

Did you know that your cat could be using its purr to manipulate you for food or attention? Dr Karen McComb of the University of Sussex researched the different sounds cats make and identified what she called a 'soliciting purr'. Unlike normal purring, it incorporates a 'cry' with a similar frequency to that of a human baby – and that makes it very difficult for humans to ignore. She believes that cats can learn to exaggerate this compelling sound when they realise it's having the desired effect on their owners.

*If we treated everyone we meet
with the same affection we bestow
upon our favorite cat, they,
too, would purr.*

Martin Buxbaum

*A cat can be trusted to purr when
she is pleased, which is more than
can be said for human beings.*

William Ralph Inge

For I will consider my Cat Jeoffry...
For in his morning orisons he loves the sun and
the sun loves him.
For he is of the tribe of Tiger.
For the Cherub Cat is a term of the Angel Tiger.
For he has the subtlety and hissing of a serpent,
which in goodness he suppresses.
For he will not do destruction, if he is well-fed,
neither will he spit without provocation.
For he purrs in thankfulness, when God tells him
he's a good Cat.

Christopher Smart, from 'Jubilate Agno'

No one shall deny me my own conclusions, nor my cat her reflective purr.

Irving Townsend

It is a very inconvenient habit of kittens (Alice had once made the remark) that, whatever you say to them, they always purr.

Lewis Carroll, *Through the Looking-Glass*

Even if you have just destroyed a Ming Vase, purr. Usually all will be forgiven.

Lenny Rubenstein

🐾 COOL CAT 🐾
Smokey

In 2011 a pet cat called Smokey from Northampton was credited with having the loudest purr in the world. The average cat's purr notches up 25 decibels, but when measured Smokey's purr was found to come in on average at a thundering 80 decibels. When measured at close range, her purr was found to reach 92 decibels – that's the same noise level as a hairdryer, a lawnmower or even, it is claimed, what you would experience watching a Boeing 737 coming in to land.

The cat's owners Ruth and Mark Adams found that Smokey's purrs were so loud that they struggled to hear the television or radio or have a telephone conversation if she began purring in the same room. The cat purrs most of the time at varying volume levels – in fact, the only time she is really quiet is when she's asleep.

The couple already owned two dogs and two other cats and adopted Smokey from the rescue centre NANNA for their ten-year-old daughter, Amy. Though twelve-year-old Smokey is by far the loudest cat they had ever heard, she soon became a much-loved family member.

At the time of writing, the Adams had submitted an application to Guinness World Records in the hope of gaining the official title of 'Cat with the Loudest Purr' for Smokey. The Guinness World Record for the loudest scream by a human is 129 decibels; for Smokey to enter the book of records with a purr measuring over 80 decibels would indeed be an astonishing feat.

Catnapping

*A little drowsing cat is an image
of perfect beatitude.*

Jules Champfleury

The sleep of the cat, though generally very light, is, however, sometimes so profound, that the animal requires to be shaken pretty briskly before it can be awakened. This particularity takes place chiefly in the depth of winter, and especially on the approach of snowy weather. At such periods also, as well as some others, the animal diffuses a fragrant smell somewhat like that of cloves.

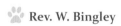 **Rev. W. Bingley**

To Sleep, Perchance to Dream

One thing all cats have in common is the enormous amount of time they spend asleep. Because they are predatory meat-eaters, cats do not need to spend a lot of time eating. To conserve energy in between meals ready for the next hunt, they spend their time sleeping in warm, comfortable places. They usually tend to nap, but if relaxed enough to enter deep sleep it has been recorded that they do produce the same brainwave patterns as humans when dreaming, and their bodies twitch. It's easy to imagine that they are dreaming about the day's activities. Unlike humans, while asleep a cat's hearing becomes even more acute than when awake, to provide warning of danger.

You can not look at a sleeping cat and feel tense.

Jane Pauley

One of the ways in which cats show happiness is by sleeping.

Cleveland Amory

A drawer, it chanc'd, at bottom lin'd
With linen of the softest kind,
With such as merchants introduce
From India, for the ladies' use–
A drawer impending o'er the rest,
Half-open in the topmost chest,
Of depth enough, and none to spare,
Invited her to slumber there;
Puss with delight beyond expression
Survey'd the scene, and took possession.
Recumbent at her ease ere long,
And lull'd by her own humdrum song,
She left the cares of life behind,
And slept as she would sleep her last...

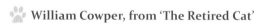 **William Cowper, from 'The Retired Cat'**

Sleeping together is a euphemism for people, but tantamount to marriage with cats.

Marge Piercy

Cats are rather delicate creatures... but I never heard of one who suffered from insomnia.

Joseph Wood Krutch

Farewell, Feline Friends

No amount of time can erase the memory of a good cat.

Leo Dworken

A Ripe Old Age

Domestic cats reach adolescence by just six months, and by the time they are a year old they are considered fully grown. Around the age of nine or ten they are middle-aged, and those that reach their twenties are feline centenarians. A mixed breed moggy will tend to live longer than a pedigree puss, and females often outlive males by up to two years. Neutered and spayed cats also seem to have an extended life – perhaps because they live life at a calmer pace without the stresses of parenthood and territorial battles. Improved diet and veterinary care have increased the life expectancy of cats from 12–15 years to 16–20 years for a healthy cat, with a few exceptions hanging on in by their claws for even longer:

- A cat named Flook from Cumbria received a telegraph from the Queen on the occasion of his twenty-third birthday – his hundredth birthday working in cat years!

- The oldest cat recorded was Creme Puff from Austin, Texas. Born on 3 August 1967, she lived an impressive 38 years and 3 days until her death on 6 August 2005.

- In 2011, a cat named Lucy from South Wales became a new contender for the title of oldest cat. If the year of her birth (given as 1972 by her owners) could be verified that would make her thirty-nine years old – twice the age of the average cat.

PS

Postscript from a letter from Charlotte Brontë
to her friend Ellen Nussey:
PS – Also, little black Tom is dead. Every cup,
however sweet, has its drop of bitterness in it.
Probably you will be at a loss to ascertain the
identity of black Tom, but, do not fret about
it. I'll tell you when you come.

We mourn the loss of our little pet,
  And sigh o'er her hapless fate,
For never more by the fire she'll sit,
Nor play by the old green gate.

The little grave where her infant sleeps
Is 'neath the chestnut tree.
But o'er her grave we may not weep,
We know not where it may be.
Her empty bed, her idle ball,
Will never see her more;
No gentle tap, no loving purr
Is heard at the parlor door.
Another cat comes after her mice,
A cat with a dirty face,
But she does not hunt as our darling did,
Nor play with her airy grace.

Her stealthy paws tread the very hall
Where Snowball used to play,
But she only spits at the dogs our pet
So gallantly drove away.

She is useful and mild, and does her best,
But she is not fair to see,
And we cannot give her your place dear,
Nor worship her as we worship thee.

**Louisa May Alcott, 'A Lament (For S. B. Pat Paw)'
from *Little Women***

One cat just leads to another.

Ernest Hemingway